If one is lucky,

a solitary fantasy,

can totally transform

one million realities.

- Maya Angelou

A

QUEST
FOR
QUINTESSENTIAL
OPRAH

Symbiosis:
Earth Energy

Valerie Bline

ISBN-10: 0692215255
ISBN-13: 978-0692215258

Dedication

The plan is to honor everyone who despite all odds, has never quit seeking justice for all. The ones who believe that ultimately, humanity can rise from the ashes of despair to become a good force worth saving. I offer these brilliant words:

"I love those who can smile in trouble

who can gather strength from distress,

and grow brave by reflection.

'Tis the business of small minds to shrink.

But those whose heart is firm,

and whose conscience approves their conduct,

will pursue their principles till death."

- Leonardo Da Vinci

"I hate a song that makes you think you are not any good. I hate a song that makes you think that you are just born to lose. Bound to lose. No good to nobody. No good for nothing. Because you are too old or too young or too fat or too slim. Too ugly or too this or too that. Songs that run you down or poke fun at you on account of your bad luck or hard traveling. *I'm out to fight those songs to my very last breath of air and my last drop of blood. I am out to sing songs that will prove to you that this is your world and that if it has hit you pretty hard and knocked you for a dozen loops, no matter what color, what size you are, how you are built, I am out to sing the songs that make you take pride in yourself and in your work. And the songs that I sing are made up for the most part by all sorts of folks just about like you.*"

- Woody Guthrie

To you, keeper of our dreams, please embrace this vision as a portal to be truly seen and heard. A way to support and illumine your efforts and accomplishments. A way to tear down some walls and build a courageous, inclusive new reality. Your sacrifice is acknowledged and appreciated. Thank you for your tenacity.

Introduction

We are all trapped in a real life global scenario of "Horton Hears a Who". This is happening simultaneously as the Lorax just stops the Once-ler from cutting down the last truffula tree. Rather than imagining some Whos on a speck, whose only recourse is to make a loud noise. Close your eyes and see a boat in the vastest ocean. The boat is filled with extraordinary people who not only need to be heard because they are there, but visionaries who must be seen because they are making a difference. As the vessel gets deluged by storms and surrounded by massive debris which shatters the hull and shreds the sail, the crew reinvents the ship over and over. They work together with uncompromising resilience and good humor to salvage the ship from doom. Their efforts create a safe comfortable vessel, better than new. As soon as word of their heroic efforts reaches shore, there is a new disaster. The people on land are told: "pay no attention to those people, their ways do not work, we will handle everything, The Earth will be fine". And they are believed.

Right now it feels as if there is only one teeny tiny boat out there, without direction or hope. When in reality, there are fleets of ships, their sails filled with wind and many in the shipyard being built by strong hands and open hearts. They are in *every* community in America, salvaging what works, creating new technology, and they are forging paths that rebuild community and restore connections. In your hands is a way to give them a voice and visibility so everyone in the world can know, for certain, that they do not act alone and people can learn the truth of their efforts.

You are holding fertile loam for the seeds of movements. It unites the "Us's" Harvey Milk spoke of, who struggle to move one step ahead and get pushed three steps back. It does have a show at its core that will serve as a hub. The spokes will be forged of other media and technology. It is time to question the actual benefits of our current technology. What if its primary intention honestly served us to make our lives better and preserve our only home as a crucial goal?

You are holding lessons about relationships and connections that some will discover for the first time. This goes beyond learning it will show and at times initiate the doing. Real issues are addressed and followed up after the episodes through other media and technology, with links to individuals, local groups and resources.

You are holding a force that will change lives in substantive ways, not green washing or superficial makeovers. The standards will be high, but we possess an abundance of talent to meet any challenges. People of all ages, from all groups, neighborhoods and stations of life are compelled to engage with the show and its extensions.

What you are holding in presentation isn't glitzy or in television format. What it *is*, is honest, from my heart and my deepest being. Throughout I have culled gems of wisdom from the ethos, the collection of these words alone makes it a worthwhile read. If my efforts to inspire are successful, please eliminate the few degrees of separation and deliver this to Oprah's hands. If you love but one child who will live on this Earth after you leave it, you ought to honor this Quest. Then I promise, music, art, laughter, science, poetry and love. These things we need most to save, will surely save us.

THERE IS A CRACK IN EVERY THING.
THAT'S HOW THE LIGHT GETS IN.

- Leonard Cohen

The salvation of this human world lies nowhere else
than in the human heart,
the human power to reflect,
in human meekness,
and in human responsibility.

-Vaclav Havel

One does not discover new lands

without consenting to lose sight of the shore

for a very long time. — *Andre Gide*

October 2014

Dearest Oprah:

With this rare opportunity, I would like to say how sorry I am for your loss of the incomparable presence in your life who will always be your Maya Angelou. Without personally knowing her, with only knowing her virtual hugs, we will miss the love she gave so freely that we all knew she spoke directly to us. We must be grateful for how her mastery of words bequeathed us wisdom to endure and courage to flourish.

I often wonder why humans trap themselves in boxes and force themselves to live in a linear system which does not exist in Nature. To sacrifice the twists and turns of our journey for a fragile certainty seems a waste of potential joy.

From the first line I penned for your eyes there has been this badgering inner voice: "hurry, hurry, the moment is passing". I am glad I silenced its persistence. Now, I am grateful for every delay till this moment. Each delay has contributed knowledge and experience. It is the perfect time and place for you to ponder my huge request.

You and your network, blessed with the best team on television are well positioned for an unprecedented action. Though The Earth has multitudes of advocates around the globe, their resilient voices are stifled by the powerful few. I have a request on behalf of The Earth.

As someone with chutzpah enough to appoint myself spokesperson for The Earth, I implore you to embrace the most important and extraordinary teaching moment of our time with wild abandon. The lives of children today and tomorrow depend on all of us being exceptional students and teachers. I am sending you this proposal for a series that will touch all corners of The Earth through TV and the Net.

From a place of joy and wish for peace, my request of you, is to expand the collective journey you live with the world by immersion into this inclusive force as diverse as Nature herself. It is born of creativity and interactive connections with and reaching far beyond the series. It will reawaken our true connections to each other and our terrestrial home. This collaboration and sharing will occupy a most sacred space for humanity's future.

The voice you lend to The Earth will be a flourish of sweet trumpets for all the world to hear. People will finally understand, OUR ACTIONS TODAY MATTER. Future generations, our children depend on The Earth for a secure haven where their dreams will dwell. I wish you would grant The Earth fruition of my proposal. I promise you peace, presence, love, outstanding progress, feats of energy, laughter and fun.

humbly yours,

Valerie Bline

Valerie Bline (self- appointed spokesperson for The Earth)

p.s. (because editing brings more thoughts):

The perfect storm of inhumanity revealed by Hurricane Katrina, temporarily shocked our nation and the world. Among all the exposed injustices climate change is the most urgent and crucial. Al Gore's pairing of inconvenient and truth, precisely paraphrased our collective inaction. The choir listens and takes action, while mainstream America, led by the corporate-driven media is mired in misinformation and fear.

When you used your show to remind America that people in the Gulf were still struggling, your passion was so palpable, I immediately knew you would help. My research and writing began. When you brought "A New Earth" to us and dared to leave the sanctuary of the network behind, I knew for sure, when this idea reaches you, you will welcome the grand possibilities.

Through the years you have been brave to let us witness your personal and very public metamorphosis. I believe you care deeply about the world you leave behind. I keep saying, this work you hold is NOT a mere pitch, it is instead an essential plea for a venue where we can all envision and create a path for a future satiated in peace. I believe you will not hesitate to do this greatest mitzvah for the sake of generations to come.

"My soul can find no staircase to heaven unless, it be through Earth's loveliness."

- Michelangelo

Table of Contents

ONE

Men occasionally stumble over the truth,
but most of them pick themselves up
and hurry off as if nothing has happened.

-Winston Churchill

With mass media as a conduit, the fossil fuel industry tells us that solar, wind, geothermal and other renewable resources are too costly and unproven. Yet every day, The Earth is showered with 35,000 times more solar energy than we consume in one year. People and animals die from toxic air and water, the oceans are depleted of life, the planet's temperature rises and we have even *changed the course of the wind* which our children will inherit. All this in the name of "cheap" energy. We deserve the truth.

The truth is power generation is as simple as turning a turbine. An easy job for wind, flowing water or by boiling liquid as most "conventional" power plants do now. The truth is, in the U.S. and around the globe, fiscally viable, sustainable, clean energy powers the world, now.

The truth is clean coal does NOT exist. The premise is built on a notion of storing carbon underground, "carbon capture" or sequestration requires burning more coal in the process. It gives us more carbon to store.

The truth is hydraulic fracturing and oil sands production are irreversible environmental disasters.

The truth is NOW, in Spain, they can store solar energy underground for use at night. The truth is wind energy can also be stored underground.

The truth is NOW, we can build clean homes that are carbon neutral with walls that sequester carbon in the process. That's true carbon capturing!

The truth is if we stay this course, science shows the oceans are in grave peril. The truth is we need the oceans to survive.

The truth is clean water and air are finite and not an optional commodity for life. The truth is time is up. WE, are on a precipice.

The truth is The Earth, will flourish without humans.

TWO

OUR LIVES BEGIN TO END THE DAY
WE BECOME SILENT ABOUT THINGS THAT MATTER

-DR. MARTIN LUTHER KING JR.

September 27, 2012 marked the 50th anniversary of "Silent Spring". What Rachel Carson courageously taught the public, at great personal expense, saved countless birds, animals and humans from irreversible harm. Over fifty years later, the same industrial giants have been joined by myriads of others to pillage The Earth for profit at the expense of all life.

In the first chapter of "A New Earth", Eckhart Tolle explains how the advances of war technologies over a short span of time intensified to accelerate man's ability to kill and destroy. That increased efficiency connected less people directly to the actual act of taking lives. In the same way, technology has fast-tracked the plundering of Earth's resources. The casualties from irreverent actions against life on The Earth are increasing at a lightening pace with a more destructive end. New chemicals pose more pervasive yet ambiguous threats. Used globally, neonicotinoids accumulate by small doses (like DDT) and destroy massive bee populations and water supplies for everyone. Since 70% of our food supply is pollinated by these selfless creatures, the loss of these bees must be stopped now for our survival.

Topsoil is a complex entity, teeming with life on an easily visible and a microscopic scale. Third in line to clean air and water, fertile soil is the most essential component of The Earth's thriving diversity. Our rapacious, intensive farming practices are a suicidal denial of the truth. At current depletion rates, The Earth will be void of arable topsoil in sixty years. At the time Rachel Carson was sounding the alarm, corporations declared chemical warfare on all bugs. Despite poisons and altering Nature's designs (GMOs) they still reproduce and adapt faster than we can find new ways to kill them.

We commoditize everything and fancy ourselves as "producers" of food as if seeds were our invention and Nature has nothing to contribute. In her book "Biomimicry", Janine Benyus stands in a Minnesota wheat field and observes: "nothing extraneous is *allowed* to grow here: everything has been stripped down to its least diverse form". For over 3.8 billion years, Nature has shown how essential diversity is to all life. Wouldn't it be smarter to cultivate cooperatively rather than our feeble attempts to conquer Nature?

How we treat our only home correlates directly to how we treat each other. It is all about connection, multiplied by billions, the choices we make every day have a huge effect on The Earth. Nature is a life sustaining entity with rules that do not differentiate between humans and other creatures. Every animal, insect and microbe work together as a whole system with rules. Our hubris toward these rules and the laws of physics, create a dysfunctional divide between man and The Earth.

Growing lists of extinct species, films of drowning polar bears, vanishing bees and dead birds don't deter humans from their egos' desire for more. When will the impact from "natural disasters" dominate in America enough to matter? To care? To change?

We've always been connected. Now, because of over population and obscene global market demands put on The Earth, we need to accept real boundaries. If we do not change course rapidly, on the 100th anniversary of "Silent Spring", our children and their children will look back at our generation and ask why we didn't change when we were thoroughly warned. Why the oceans are dead? Why people die for lack of water while others seek dry ground? Why did we lack the courage and will to leave our comfort zone, when we were fully aware of the consequences? What were **WE** doing? What will our children endure for our hubris? How could a people, so technologically brilliant not learn the meaning of **one word: ENOUGH?**

THREE

Let us not look back in anger or forward in fear,
But around in awareness

– James Thurber

It is the height of irony to consider that the main sources of current information are commonly referred to as "news outlets". It does conjure up mental images of large retailers marketing stories that they select. Then they sell them to the masses as if satisfying consumer demand. These outlets are thorough when ensuring audiences don't miss a celebrity scandal, horrific murder, epic disaster or political misstep. The coverage of one story is exploited for weeks until we lose sight of the significant events that actually change laws and affect the lives of our children.

With the risk of being judged as over dramatic, it should be considered a crime against humanity for how few know the names James Balog and President Mohamed Nasheed. In a world where the word hero is assigned frivolously, these two world citizens have heroically risked their lives to create global awareness. They have journeyed from sea floor to summit and still remain invisible messengers. Like Mark Ruffalo, they cite deep love and concern for the future of their children as their prime motivation.

The film "Chasing Ice", is a photographic odyssey of James Balog capturing historic losses with the genius of his lens. The undeniable evidence of shrinking glaciers and the inevitable repercussions are almost beyond human comprehension. We see feats of engineering and physical daring to enable over twenty five cameras to be installed and maintained in some of the harshest conditions on the planet. These cameras empower us to see glaciers disappearing at a rate that renders the expression "glacial pace" obsolete, unless the speed you wish to express is fast. What melted in one hundred years now takes a precipitous ten. The images are visually stunning in their beauty and terrifying in the truth they tell.

A dedicated staff endure brutal, Arctic conditions, traversing ice and rock, where one misstep could cost a life. They are led by James Balog with knees that seem to lose cartilage with every step and whose determination to be heard becomes more vigilant as the ice recedes further. They return with thousands of photos which are painstakingly analyzed and formatted to update the devastation. Balog then spends months on the road, obligated by what he has seen to elevate awareness and inspire action. "Chasing Ice" was not embellished by special effects, the graphic images we see are real. It was released in theaters in 2012. Two years later, our failure to pay attention to this remarkable work is shocking. I liken James Balog to Sisyphus, everyday pushing the boulder up that steep hill, again and again. Balog does it with deep conviction and the belief that his success will free his daughters from a struggle with an immovable boulder.

After being held and tortured as a political prisoner by the oppressive regime he had known since birth, and forced into exile, Mohamed Nasheed returned home as the first democratically elected president in Maldivian history. For an island nation struggling with threats from rising sea levels, the election of a marine scientist and environmental activist as president in 2008 was fortuitous.

The Maldives is a nation comprised of 1200 islands in the Indian Ocean whose very existence is in question due to climate change. It has some of the most beautiful, fragile and essential ecosystems on Earth. For over 3,000 years, the Maldivians have flourished on the two hundred habitable islands. Projected sea level rise will submerge this nation, the effects are already felt in places where sea water has infiltrated the supply of potable ground water. These Islanders are third on a list of imminent climate refugees. The Maldives will soon be a homeless nation, without a country if we continue our unconscious behavior.

After Mohamed Nasheed took office, he boldly declared that his country would achieve carbon neutrality within ten years and implemented a plan. What a noble goal for a nation encumbered by climate change despite its inconsequential footprint. He believes that international decisions are made due to small decisions, every solar panel counts. Nasheed champions clean energy production to advance developing countries. Prosperity and survival await those who embrace proven alternative energy production and shed destructive fossil fuels. History was made when Nasheed held a Cabinet meeting under water to stage the real life drama his country faces.

In direct contradiction of his personal diminutive stature and that of his nation, President Mohamed Nasheed unleashed a whirlwind to campaign for humanity at the 2009 Copenhagen Climate Conference. This lone gadfly, deftly confronted world leaders. They had come prepared to agree to disagree, shirk their obligations and avoid political consequences. Now, they were in negotiations with a man who lived for 18 months in solitary confinement on a sweltering beach, and his resolve was as strong as the steel box that held his body baking in the sun. He endured beatings and separation from his family and home. Mohamed Nasheed has no hesitation about seizing every moment. He knows of loss and he has lived to tell of the high stakes. This President of a tiny island nation put it all on the line and extracted the most comprehensive climate agreement in history. Though a non-binding document, it contains the now familiar threshold for carbon levels of 350 ppm, a critical mark which we have since surpassed. Nasheed, even after being removed from office by the old regime at gun point, in a quiet "coup d'état", still holds fast to a relentless optimism.

Despite death threats, he is still unflappable. He will do whatever it takes to save The Maldives and restore democracy. Before Copenhagen, he addressed the U.N. in New York. In his plea for survival he spoke: **"... we continue to shout, even though, deep down, you're not really listening"**. A glance around the room reveals the deafness. Nasheed believes mistrust is what holds us back. In these conflicts that make zero sense, if we do not act together, we will share a common demise. It is a theme in science fiction, that if Earth were invaded, man would unite for survival. This parable is less ridiculous than the reality of our deafness. We need to hear the Maldivians shouting for compassion, shouting to live. We need to answer: "YES, you matter and we will help".

If I am not for myself, then who will be for me?

And if I am only for myself, then what am I?

And if not now, when?

-Hillel The Elder

FOUR

There are down sides to everything; there are unintended
consequences to everything. The most corrosive piece of technology
that I've ever seen is called television. But then again, television,
at its best is magnificent.   -Steve Jobs

Extreme Makeover: Home Edition, took their act to Pinon, Arizona and unwittingly chose corrosive television over magnificent. Maybe the actual show was just average for the nature of the series. In their basic scenario each week, they would renovate or build a home for a deserving family facing difficult life circumstances. They were the superheroes of home makeovers, each episode replete with heart tugging moments and tears all around.

The impetus for this episode was an invention of Garret Yazzie, a 13 year old Navajo and his family. Garret, one year earlier, designed and built a solar heater for his family's dilapidated trailer. Without formal training or skills, a 12 year old kid, combined soda cans, Plexiglas and junk car parts in a box to harness the sun. Garret's family was warm while his asthmatic sister breathed easily without smoke from wood and coal. His innovation yielded vast improvement for his family.

Combining an alternative energy theme with Navajo Hogan building design, they took the Reservation by storm. With a big budget show, the Yazzies were presented with a spacious Navajo inspired home powered by wind and sun. What the viewers did not see, was that the family was left with an unaffordable nightmare. In their haste, the building design and systems were incorrectly executed. This huge, poorly insulated house, sent them back to burning wood and coal for warmth. This house was too big for the solar heater from the little trailer. Garret's sister had to move in with relatives because of her asthma.

Mrs. Yazzie contacted those involved who refused to fix the problems, only one contractor helped the family resulting in a minor difference. At one point, the Yazzies had to explain in the press that they were not being ungrateful. Sadly, the one year warranty expired and they were left with expenses above their means.

This is an inexcusable, epic waste of a tremendous teaching moment for America on a major network, in prime time. Born of goodwill, this project proved to be a disaster not because of faulty technology but due to haste, incompetence and apathy. The house should have been a triumph for alternative energy. It should have shone a light on social, educational, spiritual and economic issues. Much potential and many, many opportunities were squandered. Garret's family belongs to the Coyote Pass Clan, in their culture, coyotes are messengers. At 13 years old, this courageous coyote, delivered a message to us for our future and for the future of his children. We have failed him and we have failed ourselves.

Fortunately, failure is a key component of success. The other good thing about failure is that it only happens as a result of true effort. When we give a hand up to the Garret Yazzies of our world, we give courage so they can dare try and succeed again.

FIVE

SYMBIOSIS: Earth Energy

If you want to build a ship, don't drum up people together to

collect wood and don't assign them tasks and work,

but rather teach them to long for

the endless immensity

of the sea.

-Antoine de Saint –Expurey

Our connection to The Earth and to each other is severed and hemorrhaging. Mountains of evidence prove that our survival as a species on this planet is at stake. Our only chance, is in raising awareness, understanding and motivation, a serious change in the status quo. Time is of the essence. Swift bonds must be forged, truly connecting human to human and humans to all life. We must empower each other to do, we must present them with wings of opportunity. On with the show:

THE

MOST

IMPORTANT

TEACHING

MOMENT

EVER

SIX

SYMBIOSIS: Earth Energy

It feels like the void only Oprah Winfrey can fill effectively. Each day we wake to face the world. We are bombarded and overwhelmed by growing, insurmountable problems. At stake is the essential habitability of our only home, The Earth, the place where all human life happens. Without this safe haven, all the efforts to make lives better will be wasted. Now is the moment people of the world are primed to truly **SEE THE EARTH**.

Energy is the esoteric and physical link common to everything. This proposal offers hopeful answers to unite souls and create what Chris Hayes called a "collective will" to preserve a livable Earth from destructive humans.

In thirty years on television, no one has inspired us to "long for the sea" with a "collective will" to be our best, better than Oprah Winfrey. Her desire to connect people to The Earth and each other has raised the consciousness of her audience in powerful, life changing ways.

Unlike many with international celebrity status who squander the opportunity to be a positive force in the world, Oprah has always used her well earned respect and her heart beyond any fiscal collateral to make lives better many times over. Since it is here in America, that we lead the world on a destructive path, it is here where the learning must begin with our most renowned teacher.

Oprah's broad demographic transcendence is vital here in the U.S. To foist the future of our children and their Earth on Oprah's full agenda seems a bit unfair. But who else can speak to a struggling Mom in the South Bronx, a diehard NASCAR fan in the Midwest and everybody in between? Who else could have over half a million on computers from around the globe, at insane hours, to share "A New Earth" experience with Eckhart Tolle? Every week for ten weeks! Phenomenal.

The education and empowerment of all citizens, especially children, with Symbiosis is ground breaking. We witness the creation of new, humble, leaders to reconnect us to The Earth, our neighbors and our global community. We see the spirit in the philosophy of Ubuntu (Xhosa wisdom: I am because we are) applied to business and community groups across the country and world.

Oprah through her school and millions of other ways, already nurtures the children who will inherit this Earth. *Symbiosis* positions Oprah Winfrey at the pinnacle of the most important collaboration (it must be an all in effort) of human spirit focused toward true resolutions that will rescue our only home for our children. She has amassed the human capital, Ms. Winfrey will prove all cynics wrong as giants clamor for chances to help.

I have no delusions about the scope of this big ask for *Symbiosis*. Nor do I believe Oprah has magical powers. I do recognize the generations around the globe whose bonds of strong love and respect for her amount to the very definition of unique. Oprah is that connective force the world needs now. Together we can achieve the Dalai Lama's most important meditation for today:

"critical thinking, followed by action".

SEVEN

PLEASE

EXCUSE

THIS INTERUPTION:

Like most of you, I consider this an unwanted foray into honest gloom and doom

(pssst.... Those of you in the choir with me are probably thinking, it's about time!)

it will be brief, but I assure you it is vitally necessary.

THE EARTH IS ON FIRE!

This is an actual fact, really hot, inferno type fires from many sources. They are compounded by the fires lit with every issue in the human sphere. Pick any topic, they are all tinder. They rage uncontrolled and cause us to build walls between and against ourselves. Our shared recklessness will ultimately destroy the only home anyone has or will ever know, The Earth.

The horrific damage done to The Earth by humans to destroy Nature, people, magnificent cute cuddly animals, irreplaceable sea life and the essentials that we need to survive, if showcased, could drive us to despair. I am not advocating ignorance of reality, at times unsettling things will *need* to be seen.

We all have different capacities to process bad things and a conscious effort is necessary to insure a balance between informing and overload. Life today is so complicated that many good causes lose support from people overwhelmed with issues which make them feel helpless.

WE CAN QUENCH THE FIRES!

Symbiosis will plant deep roots with live hearts and hands,

with clear purpose and indomitable spirits.

WE can connect,

leap from the box,

and

build

our best tomorrow now.

EIGHT

SYMBIOSIS: Earth Energy

The intention of Symbiosis is to entertain and educate in a positive, interactive way. Accurate scientific information is presented in a format that can be understood by a layman and appreciated by those with expertise.

The series approaches complex issues and restores connections with Nature. We focus on building connections to mitigate the effects of climate change for the next generations. They are our babies, our students, our new workers and parents. The show contains features which include and attract viewers from every demographic in the U.S. and anywhere in the world OWN can reach.

Symbiosis builds an on air and online community that for the first time in television history, creates a long term sustainable, active presence in actual bricks and mortar communities. Symbiosis effects positive change toward a viable ecosystem. We hold one truth sacred: *breathable air and potable water are necessary to sustain life on Earth and EVERY LIVING BEING IS ENTITLED TO THAT SUSTENANCE, FREELY AND UNCONDITIONALLY.* Symbiosis presents and implements fiscally feasible and attainable systems in actual places with real people.

Symbiosis uses teaching moments from everyday life and forges a path beyond superficial solutions, far beyond recycling. Action creates local organizations and local sustainable businesses to reconnect us to Nature and each other in profound productive ways. Viewers become students who already know in their hearts that we all need a new direction where we all take serious ownership of our actions. Symbiosis provides steady reassurance for what is precious and essential. Symbiosis gives voice to existing groups which empower the seemingly powerless and acts as a conduit connecting humans in groundbreaking ways so they can change their stars.

Symbiosis illuminates the big picture, every action in the world affects someone else. It is the "butterfly effect" on steroids, what we do now, each day, saves a species from extinction or The Maldivians from homelessness. *Symbiosis* offers hope that our "collective wills" make righteous choices because we discover, for certain, our lives depend on honoring and cooperating with all life on Earth. We all must live by grace. If grace can be taught, *Symbiosis* will be that teacher.

NINE

the World is all gates,

all opportunities,

strings of tension

waiting to be struck.

-Ralph Waldo Emerson

Symbiosis opens all gates wide to change perceptions conceived of fear and ignorance. Television, at its finest, played a quiet role to advance the Civil Rights Movement. It brought talent like Nat King Cole, Harry Belafonte, Ella Fitzgerald, Sammy Davis Jr., Lena Horne, Sidney Poitier, Bill Cosby, Diahann Carroll and Redd Foxx into American living rooms. It granted a frightened, misinformed society a new perspective. Television crafted a comfortable familiarity to initiate acceptance. Incredibly, in 1977, a nation sat riveted to their televisions for an unprecedented eight nights captivated by Alex Halley's, "Roots". Though not a panacea, these new portals for a nation divided, were the most important accomplishment of television until Symbiosis.

Symbiosis respectfully and intelligently harnesses the talent and power of celebrities from all realms to calm unjustified fears and expose misconceptions of viewers. Through the magic of television and other media, Symbiosis provides opportunities to see energy systems we've been told are unproven that already provide affordable power for people they revere. Those celebrities show their adoring public other locations where opportunities abound for everyday citizens like them. Symbiosis provides knowledge that will place actual power in the hands of viewers for the first time.

Symbiosis utilizes celebrity wisely to create positive inroads to reclaim our connection to Nature and each other. Famous humans from all venues seize the chance to teach in unique ways of their choosing. Their passions are alive as they reveal their conscious life choices. Symbiosis was the first to ask or care about these important decisions in their lives and they relish the opportunity. We revel in their honest enthusiasm about The Earth, what they think the future holds and the ways they dedicate their time to help. Through Symbiosis, their talents become tools to further their commitments to the next generations. These heartfelt interludes will be artistic, funny, musical, dramatic and best of all, inclusive, compelling television at its finest.

TEN

I think a hero

is any person

really intent

on making this a better place

for all people. – Maya Angelou

People *really intent* on making The Earth a flourishing, sustainable planet come in all shapes, sizes, colors, nationalities, religions and other fictitious differences that can blind us. Everyday around the world, they work tirelessly against impossible odds. They offer a hand up, wade through sludge and red tape. They sacrifice time with loved ones and money. They endure bad weather and criticism from the uninformed and frightened and most times they smile or even sing while they do whatever it takes. They persist when the world rails against them until circumstances make the truth clear and they keep their eyes on the prize long after. When asked why he added his voice to the fray, Mark Ruffalo put it succinctly, he stated simply: "because I love my children".

These heroes are all around us. Symbiosis lends them a spotlight and a giant megaphone as a positive force to teach by example. Regular people and celebrities show how many actions, small and large impact our quality of life and our ultimate fate. Symbiosis brings awareness of our responsibility to each other, all our choices are consequential for our friends, neighbors, children and those we'll never meet. By considering others before they act, these heroes are living examples that illustrate the true meaning of "personal responsibility".

To initially attract viewers, Symbiosis accepts the generosity of celebrities who share their choices for a better world. After they've tuned in, viewers are compelled to become loyal students when they see their neighbors and children involved and making a real difference. They find a new truth, despite what they were told, it is not too late and we all have the power to do good. Symbiosis provides all Harvey Milk requested: "give 'em hope". Symbiosis extends far beyond hope, Symbiosis provides The Earth and humans, solid, tangible answers, results and a clear way forward into a bright future.

ELEVEN

THIS MACHINE SURROUNDS HATE AND FORCES IT TO SURRENDER

-Pete Seeger

The words above encircle the face of his signature instrument, the 5-String Banjo. They embody a man who for seven decades, wielded the power of music. Pete Seeger inspired and effected more good as an advocate for humans and The Earth than possibly any one ordinary soul. His faith in people to eventually do the right thing was relentless, till his last breath. Whether they know his name or not, anyone who has raised their voice in song for justice anywhere in the world likely owes a few notes to Pete.

As a perennial force for justice, he lent his voice and presence to help many causes. On October 22, 2011, at age 92, he marched thirty six blocks, on a cold night, to rally with Occupy Wall Street in Columbus Circle. There, he led us in "Climbing Jacob's Ladder", his voice and feet made news across oceans. When the McCarthy blacklist barred him from concert halls, Pete nurtured America's youth from elementary schools to universities. Returning full circle, in his last years he spent much time with children in Beacon Elementary. Pete has made great progress, his light shines in their eyes as they raise their voices with love.

This power of music is an integral, vital element of Symbiosis. The theme of Symbiosis is a collaboration, the song has become a part of our fabric already. Every week there is a renowned guest musician. We meet new voices, without limits these upstarts know they are a force for humanity's future. Much of their music becomes sacred gospel for The Earth and clear instruments of reckoning.

Viewers contribute original music which brings their people to Symbiosis. Upon watching, they discover *they* have skin in the game beyond their friend or relative.

From the first minstrel, music has always carried our narrative. As Pete said so clearly: "The key to the future of the world, is finding optimistic stories and letting them be known". With this optimistic inspiration, the more we know, the more we can do.

TWELVE

The purpose of art
IS TO LAY BARE THE QUESTIONS
WHICH HAVE BEEN HIDDEN
BY THE ANSWERS

- James Baldwin

Visual artists, observers and keepers of The Earth's light, have forever chronicled and honestly critiqued our world. The artist's sensitivity to their surroundings gives them unique insights and perspectives especially to reflect on the degradation of our environment, urban and rural. What would Frederic Church and other painters from the Hudson River School (mid-19th Century) think about the landscapes of America decimated in the last century? How could we explain to them why the trees didn't matter? We all bear some responsibility for the uglification of this world.

"Progress" has facilitated the unconscious destruction of The Earth, diminished our awareness and respect for life itself. It has splintered the essential connections to humans and Nature strongly grounded in and by art. Every child naturally begins life as a creator of forms and pictures reflecting life around them. The perceived monetary value of visual art cannot measure the level of awareness created by the artist or their contribution to the whole.

Visual art, once an integral part of the commons is relegated to only the richest venues and private collections. The fiscal chainsaw as applied to school budgets leaves creativity gasping for air in its dust. Beyond Pollack or Warhol, do we know or ever see America's contemporary artists? Our society invests in ephemeral decadence and more money, while the most precious things, the works and passions of children are denied a chance to grow. Imagine if Michelangelo's genius had been stifled, it would be the ultimate insult to our humanity. We desperately need these insightful keepers of light and color to guide us on how best to nurture our precious world full of souls.

Symbiosis shines with artists who bring us Earth's beauty and truth. Especially those who create art with discarded materials. Their work uses Nature's principles, where nothing is wasted and everything nourishes. Sometimes, the featured "pre-owned" components lead us to explore the source of the materials. We experience a merge of science and art which opens doors to new industry processes that minimize waste and upcycle things.

The show especially highlights art in outdoor spaces and gets more art into public places. Our results are extraordinary, beautification for the people, by the people. Symbiosis inspires fun art, serious art, art that improves air quality, art that harnesses energy and interactive art which spills over into Indie films. Yes there will be shorts!

Symbiosis challenges and enables artists, from seedlings to those with deep roots, to live their potential. They create pieces that are displayed in their communities and on Symbiosis. This art has returned character and charm to the otherwise homogeneous Main Streets of America. Local print shops produce prints of work created by local artists. The funds raised support community projects.

Throughout history art has adorned human figures. Symbiosis inspires fashion designers to choose materials that are Earth intelligent and upcyclable. We witness and champion healthier, smarter fabrics for our lives as they are handed down to our kids.

The presence of Artists is essential for they bring unique and fervently dynamic undertakings to Symbiosis. Let the Renaissance begin anew!

THIRTEEN

It's not personal, it's just business.

- Nora Ephron (Joe Fox)

While it seems a bit strange to use a romantic comedy to make a serious case, the line: "it's not personal, it's just business", in "You've Got Mail", was used repeatedly. The actual truth of the narrative is used everywhere in real life to explain away and condone our bad behavior in our society. Joe Fox's nemesis, the delightful Kathleen Kelly, countered: "whatever anything else is, it should start by being personal". She is right, and it is time for our business model to be as nice as Kathleen Kelly. Precious life, essential life, we must consider it ahead of money lest we be thought insane enough to steal the breath from our children.

In America, we are taught and meticulously conditioned to revere the Joe Foxes and corporate giants like Fox Books. Clouded by consumerism at birth, we are all complicit in the theft of our democracy. We are humans, now reduced to consumers. Citizens without civility, demoted to taxpayers. This Machiavellian system of commerce gives business and politicians cover to disenfranchise us and pillage The Earth. In their wake, they leave polluted waterways, toxic soil and air unfit to breathe. They fracture communities and pit neighbor against neighbor. If you are a miner with black lung, or have cancer from living in the shadow of a refinery, or can't afford food or rent, you are merely "collateral damage". You, are an expense line factored into "the cost of doing business". When monetary costs in America cut into profits, the corporations outsource their pollution and death to somewhere that they are welcomed by desperate others. These "others" are our fellow human beings.

These corporations are not people, but they do provide masks of anonymity which shield actual individuals from liability. If we were all paying attention to their actions, we might not lend our support and reward them with our patronage. Our present society upholds this system that allows someone to work and live in an environment that will disable or kill them with full disclosure. Many respected "experts" tell us: "that's just the price of cheap_____ (fill in the blank)", because they have no connection to those "others".

To vilify and place blame, is divisive, counterproductive and diminishes our humanity. Wouldn't we be better if we regarded our neighbors far and near, who toil for us, as human beings, as friends? How did we become blind to the sameness of our faces and hearts? Wouldn't we be in line with our values if our neighbors could feed their family and live in a decent place? Wouldn't the quality of all our lives be enriched? Our first step could be to support corporations that treat our neighbors well. Companies led by humans who are accountable for their decisions, whose actions clearly support a deep core belief that we are all important. All our lives matter.

Industrialization has been going on some two hundred years, it seems clear the system is broken along with its promises. It is time to end private profits with public risks and other concealed public costs. There are new models, proof that companies do NOT have to plunder The Earth and exploit human beings to turn a profit. We, the people, do not have to live in indentured servitude until we are too broken to work and get tossed to the scrap heap. There are ways we can all have meaningful work, reap fiscal solvency and ensure time to live and enjoy a dignified retirement. This most basic principle of a healthy society will be restored and realized by nurturing seeds of conscious connection.

Symbiosis is a showcase for ways to rediscover this promise and to right our course. Just as there are clean ways to power the planet which have been hidden from view, there are corporate leaders and business owners, large and small, everywhere, who appreciate and respect the ones who make their lives possible. One of Symbiosis' most impactful lessons teaches us how business is done better by treading lightly on our good Earth.

FOURTEEN

No Problem Can Be Solved
From the Same Consciousness That Created It.

- Albert Einstein

If you read that we lost one of the greatest, innovative minds of our time to cancer a few years ago, your mind would probably picture Steve Jobs. Your mind would be mistaken. While we can only know him through videos from the 80s till he passed in 2011, what doing the right thing did for Ray C. Anderson, and what he did for children, is a triumph. In 1973, Ray Anderson founded a carpet company called Interface. Beginning in 1994, he took a radical and extremely brave chance to alter his company's trajectory. Interface, a global Goliath, left the path of an intensive petrochemical plunderer, to become an environmentally intelligent citizen of The Earth.

It was the power of his new creed and belief in new deeds that ignited his spirit to venture out and lead where no one had dared on such a grand scale. It should be noted that this was a major coup for a publicly held corporation. Ray Anderson, proved over and over, that "doing good, by doing right" is not only profitable but successful on all levels. On August 31, 1994, he revealed a visionary plan to begin the journey up "Mount Sustainability". When he spoke prior to that day, he appeared as a prototypical corporate head, talking of acquisitions and bottom lines.

One year later, after a just few tangible steps toward the base of the mountain, now, on a world stage, he was a transformed human being, alive with an immensely heightened purpose. He soared above and superseded profit margins and growth. Ray C. Anderson, soft spoken, straight laced businessman from Georgia took the helm in a titanic revolution on behalf of the next generations who will inherit this Earth. His new found resolve and indelible enthusiasm going forward, shined through each word he spoke and it was contagious. Till the end, this seasoned veteran of corporate America, displayed the courage and passion of a great explorer seeking a new world. Anderson charted a course for responsible industry and technology. He shared knowledge and wisdom as his sails captured the winds of a new day.

While Interface continues Anderson's efforts, there are countless graphs and figures to equally excite statisticians and environmentalists alike. The *indicators of more*, profit, sales, productivity, carbon reduction, water saved, waste reduced, air quality improved, alternative energy used and endless innovations exceed expectations. Amazingly, these numbers are not the most exciting and significant component of this new model. Pandering politicians always tout American workers as being the best, then kick them to the curb as they give Wall Street donors carte blanche. However, this retooled model led by Anderson and the staff at Interface attracts top cutting edge talent to work at, oddly enough, a carpet company. This is a dedicated company being joined by and leading others, they answer to a higher authority: KIDS! The workforce at every level from the corporate offices to the warehouse floor, earn a good living with an opportunity to protect The Earth for their children. Meaningful work and a living wage are great incentives to show up inspired. Empowering one generation to help the next by leaving them a brilliant future is an ideal America's citizens embrace. The panderers should take notice, this empowerment changes everything and this is a new day in a fair system that will enable success.

There is one more factor in the Interface equation for success that has become the proverbial ace in the hole, public goodwill. Like the esoteric nature of measuring motivation and morale, the goodwill of the marketplace does not always fit into a pie chart, yet it is a powerful force. Consumers are increasingly aware of where things originate and the impact of products on people and the environment. Whether a personal purchase or a decision made by a purchasing agent, Interface's commitment to practices that benefit the whole, is the defining factor that secures many sales. Perception is supported by real, positive actions. This is publicity that no amount of slick, expensive advertising can match. Today honest, transparent and holistic systems trump unconscious, scorched earth business models. Acts of public goodwill endure and prove the efficacy of Anderson's bold vision.

We have lost the opportunity to meet Ray Anderson, but his work lives on. His successors relentlessly continue his mission and are on track to reach the summit of Mount Sustainability by 2020. Their efforts continue forward with active resources to support and guide others. They restore and return more than they take while transforming harmful paradigms into forces for good. What an outstanding legacy!

Ironically, also in 1973, while Ray Anderson began a history of plundering, Yvon Chouinard founded Patagonia on polar opposite principles. Yvon Chouinard always believed in placing the planet and people above profits. Patagonia has treaded lightly on The Earth since inception when Chouinard introduced an obscure, untrusted piece of rock climbing equipment solely to preserve rock face. He could have grown rich quicker and avoided all the work of convincing climbers to trust their lives with something new, but his commitment to a thing then called conservation, was not negotiable. His strong ethics paid off and from the start Chouinard knew, as Anderson discovered twenty years later, doing good is good business.

Patagonia, through all of its manifestations, has learned tremendous lessons from bad situations. When they acquired a retail building in Boston, their workers experienced severe headaches and other health problems. No doubt, they could have sufficed with a mere building modification. Instead, they maximized the teaching moment. It was the intensive chemicals on the cotton clothing that contributed to the building's toxicity. The company was now obligated to consider what the effects were for all people handling and wearing the chemical laden clothing. It was the start of the organization's commitment to only sell organic cotton. Since organic cotton was not readily available, Patagonia planted seeds and prayed for rain.

One look at the Patagonia website makes clear, buying a shirt has more significance than comfort and style. Though they are keenly in sync with and take steps to care for The Earth, this is not a tiny operation in a woodland barn filled with tree huggers (not that there's anything wrong with that!). Patagonia has over 500 million in annual revenue and operates globally.

Yvon Chouinard, co-founded 1% For The Planet in 2002, which now unites 1200 other businesses to dedicate 1% of their sales to environmental non-profits. Over 700 of the tree huggers who staff Patagonia, have taken the company offer of full pay up to two months, in exchange for fulltime volunteer work with an environmental non-profit. They also consult with bluesign Systems in Switzerland, who develop green textile sciences to further shrink Patagonia's ecological footprint.

In 2012, Patagonia became a B-Corporation. B stands for benefit to workers, community and environment. This certifying body helps socially sustainable businesses to remain mission-driven as their founders intended. B-Corps now number 1,104 companies in 35 countries, spanning 121 industries.

These concepts are seeds of change in our present corporate structure growing toward human and Earth first oriented business practices. Interface and Patagonia are just the tip of the iceberg, they are working tutorials on corporate responsibility with profitability. These two elements are not mutually exclusive, in truth, they are the essence of Symbiosis. Whether we care or not, the current memes in our society condition us to say we don't care. Again, the truth is, we human beings are genetically hard wired to care, cooperate and dare I say, love.

Symbiosis will feature a company each week, so we can learn of their operation and inspiration. This will provide teaching moments in history, science, economics, and human triumphs that reconnect us. The potential lessons are infinite.

As all companies who put people before profits learn, the goodwill of the marketplace is a truly lucrative and intelligent force of attraction to potential customers. Of course this does not work if a business falsely represents itself. Symbiosis will thoroughly vet any claims made, honesty is paramount. One of the reasons our quality of life is so compromised is because we accepted too many things at face value. Symbiosis will ask better questions and with complete consideration of the big picture, we will only accept the best answers.

The children who see these exceptional companies on a tour in person or on a screen through Symbiosis are exposed to positive models of possibilities. Everyone has a chance to imagine themselves doing something good in the world while they support themselves and their families. Seeing is believing and believing that grand things will happen opens hearts and minds for a new consciousness able to brighten every neighborhood in the world.

... I've come to realize that in an organization, creativity, is unleashed by bold visionary leadership that sets aspirations so high, they take the breath away.

-Ray C. Anderson

FIFTEEN

A clever person, solves a problem.

A wise person avoids it.

-Albert Einstein

In the vastness of the universe, what we humans actually know, could be lost on the head of a pin. We need to train our focus on how Nature does everything, almost effortlessly and without waste. It would lead to intelligently conceived designs for holistic systems. It could save us from falling into the clever person's trap. We can stop spending our lives solving endless problems that we create and better spend our lives savoring the joys of life.

Symbiosis introduces the world to a new breed of thinkers who exit "the box" and then envision a world where even the box can be bodacious. One of the show's favorite scientists is Janine Benyus, a biologist who pioneers a unique, unifying scientific force called Biomimicry. This is a fortuitous trend, scientists in every field are rethinking our lives from the most intelligent perspective and reintegrating with Nature. As Benyus tells us:

"We come not to learn *about Nature* so we can circumvent or control her,

but to learn *from Nature*, so that we might fit in,

at last, and for good on The Earth from which we sprang."

Beyond the incredible things they have discovered by being humble and truly seeing Nature, these scientists apply the knowledge offered to do the most remarkable things. They prove that peace is always a wiser path than conquest.

Janine Benyus brings a radiance to Symbiosis that evokes curiosity even in the cynics. She points out that Nature has 3.8 billion years of experience in research and design perfection. We humans continue to use a flawed system which wastes 96% of resources to yield only 4% of usable product. Nature wastes zero to yield 100% gain. Maybe we could learn to change our ways of applying heat, pressure and chemicals to everything we process and find out how Nature makes ceramic with only sea water and oxygen. Benyus provides dazzling insights and reveals how Biomimicry enables vaccines to stay viable without refrigeration for transport anywhere. We see how they can harvest potable water from fog. She will introduce us to designers and architects who, with Biomimicry, work to build dwellings and cities that mimic forests to clean the air and water. Biomimicry is a wondrous source of possibilities and inspiration.

Symbiosis introduces the world to economists like Joseph Stiglitz, because he can explain fiscal issues that affect us without any secret language. He employs common sense and does not dumb down serious points. Stiglitz employs his natural wit and charm to create understanding that counters fear. What a coup, Symbiosis spawns a household name from an economist! The well of talent, knowledgeable and engagingly fun and serious people from previously untapped sources grace this world stage.

Amazing people take time from their lives to show up. They show up not just for Oprah, they show up not just because it is the place to be seen. Both reasons are true. They show up for Symbiosis, historians, poets, astronauts, farmers, chefs, doctors, heads of state, engineers, athletes, inventors, lawyers, **people who care show up because this is the grand teaching moment and they want to contribute and be a part of this movement.** People like Temple Grandin, William Kamkwamba, Wendell Berry, Severn Suzuki, Ed Begley Jr., Garret Yazzie, Jane Goodall, Jacob Barnett, Monty Roberts, Amy Goodman, Michael Pollan, Leymah Gbowee, Tom Shadyac, Chris Hadfield, Doris Kearns Goodwin, The Cousteaus, Lawrence Lessig, Margaret Atwood, Glen Keane, Robert Reich, Malala Yousafzai, Dan Barber, Raj Patel, Greg R Smith, Zbigniew Brzezinski, Vandana Shiva, Ezra Klein and JK Rowling, sign up every week. These visionaries know Symbiosis gives our children a chance for a beautiful future. Symbiosis is the vessel to right our course, so our children can find solace in the sea instead of fearing its rise. We all see, doing good for The Earth, is doing good for all, for the grand "Us". Symbiosis: Earth Energy… it is time to S:EE.

Great things are done by a series of small things

brought together.

- Vincent Van Gogh

SIXTEEN

When you look at a tree and perceive its stillness,

you become still yourself.

You connect with it on a very deep level.

You feel a oneness with whatever you perceive in and through stillness.

Feeling the oneness of yourself with all things is true love.

-Eckhart Tolle

The show opens with peace. Moments of quiet centering.

A shared silence.

It could be on location in the midst of a city, in a serene natural place, in someone's home or studio. Perhaps a darkened set, figures in silhouette while the lights come up softly to reveal the host, audience and a Qi gong master.

Viewers and live audience are requested to participate in an easy five minute Qi gong session. Qi gong is chosen due to its ease of movement, versatility and effectiveness. It can be done seated or standing, by individuals of any age or ability. Qi gong will clear old energy, increase positive energy and focus for an optimal state of understanding.

Most shows today open with loud music, an announcer, a gunshot, explosion or gruesome murder. Many thought at first Symbiosis's opening silence would be weird or a signal to procure a snack. Now, we know from viewer feedback, they love the opportunity to decompress. They've compared Symbiosis Silence to a group hug but most think of it as our collective sigh.

The show closes with thoughts, some wisdom bigger than ourselves presented by a writer, poet, child or grandparent. Perhaps a moment being in Nature doing something extraordinary. This is followed by a sacred place, breathtaking scene or work of art, ending as we all began in silence.

SEVENTEEN

In between opening and closing of Symbiosis are a broad variety of segments featuring celebrities and undiscovered people who present solutions and extraordinary ideas. Some are being used now and are viable and some are future answers to solving energy issues. Symbiosis offers new perspectives of people and places. Each day we are in direct contact with a world we never truly see. We help focus on and process what is within reach of our senses.

Symbiosis presents multidimensional aspects and talents of people and their contributions to "The Whole" beyond the measure of "human capital". It is common knowledge that human beings thrive when they find meaning, passion, creativity and purpose in what they do. Symbiosis taps into our neglected humanity.

Symbiosis transitions past the negative semantics of environmentalism by example. We transform love for The Earth, our home, into something positive and popular and present an enjoyable, educational experience. At the start, Symbiosis changed the vernacular and discovered it was unnecessary. What overcame the negativity was the joy people felt to finally see and act on opportunities that make significant progress in their lives.

There are millions of individuals, groups, small businesses and large corporations doing the right thing struggling to be heard above the din. Symbiosis shines such a bright light on their positive efforts that it becomes a giant, collective beacon and a longing to do more. We showcase companies like Patagonia, Toms Shoes, Stonyfield Farms, FLOR and others who do not exploit The Earth and humans. These corporations (fully vetted) featured on Symbiosis are rewarded with patronage born of goodwill. It is a revolution led by Symbiosis to effect a new corporate normal. This holistic approach has created a world where CEOs like Ray Anderson are empowered to act consciously to take care with The Earth and humans because it is now accepted as part of doing business. Of greater importance, this new accountability is now expected as an established cultural and societal obligation.

While money may be a necessary device in man's world to help manage our lives, it should always be a tool to serve people. The Collective Ego has twisted our values and replaced personal integrity with acquisition of financial wealth as our highest accomplishment.

Symbiosis creates a positive *EARTHCENTRIPETAL FORCE* so we can realize that this journey we live together is more valuable than our possessions. Our home and all who make it a place of love, are our true source of wealth. We treasure them. Children and their future are at the center of Symbiosis, they play an ever present integral role. Where possible, groups of students are with us on location and we link electronically to classrooms around the globe. Symbiosis is viewed at after school programs and even summer camps. The kids' participation in Symbiosis entices the current stewards of The Earth (aka adults) to action. Kid pressure is a powerful force to put adults on the straight and narrow, in some cases their concerns are the perfect nudge to make their elders finally "do something"!

While Symbiosis easily attracts seekers of knowledge, the initial hook is Society's love of celebrities. In a feature called: *Life Styles of the Rich, Conscious and Famous* on Symbiosis, we spend time with a star, musician, sports icon or other favorite person, at their home showing off their Earth intelligent life choices. In the case of Brad Pitt or Ed Norton, we meet them on their solar projects in New Orleans or Los Angeles.

Symbiosis was blessed to secure Ellen Page as our steady host. She has the perfect blend of charisma while being authentic and grounded. Symbiosis is lucky to have this brilliant, conscious woman at the helm. There is also a guest host to work with Page each week. This role on Symbiosis is more coveted than hosting Saturday Night Live in its prime.

For the first show, after Oprah's initial introduction, premiere of the theme music (a collaboration between will i am and Kenny Loggins), the first Symbiosis Silence arises in Philadelphia, where outstanding things are happening.

We begin at midfield in Eagles Stadium and it is not the green of their uniforms that made this a place to start. Lincoln Financial Field is special because it makes the world a greener place. It is energy self-sufficient, powered by an array of solar panels, vertical wind turbines, and biodiesel (from their fryer oil). The Eagles organization has proven what is possible, the elements of their bold effort have spread to over thirty other sports venues.

On the field, Ellen Page and Qi gong master are joined by Eagles owner, Jeff Lurie, some players and several classes. The silence begins and ends followed by a Qi gong session. We speak with Jeff Lurie about his philosophy of giving back to his community and the planet. With sixty one percent of Americans who follow sports dwarfing the thirteen percent who follow science, we are fortunate that Lurie uses his influence so effectively. His good work has eliminated 481 million tons of greenhouse gas from the atmosphere every year. The spread of this success will help inspire other fans whose teams now set their course for true green.

As Oprah did with Dr. Oz and others, Symbiosis sees its first celebrity on the science front as Neil De Grasse Tyson, an astrophysicist, who takes us for a tour of the stadium. We all follow him through the huge on field screens. We learn up close about the incredible, dependable power provided by the sun and delivered by the solar panels. After thirty years of innovation stagnation, technological development of solar energy has made major leaps in increased efficiency and fiscal viability. Eagles stadium when not using the power provided by the sun, feeds it to the grid for the community, essentially becoming a small, clean power plant. TOUCHDOWN!

While in Philadelphia, Symbiosis visits a former galvanized steel plant. Once deemed an unusable brownfield on East Cumberland Street, it is now the center of a sustainable urban farming operation. Green Grows offers fresh local produce at Camden Neighborhood Markets, providing a food oasis in an otherwise "food desert". Their organization helps establish roof top gardens which also help in storm drainage. They provide a CSA program and food safety training with certification for job seekers. At the markets there is free nutrition education, health and wellness checkups with doctors and help to apply for food assistance. By bringing gems like Green Grows and Farmer's Markets in the public eye, Symbiosis can increase the good they do and will help propagate their models with positive exposure.

The most important, intentional obligation of Symbiosis is to reconnect all people to The Earth. There is nothing more basic than knowing where food really comes from and no better lesson than growing our food locally. Learning how to grow what sustains us is a crucial lesson of and for life. In America, on average, food is transported by truck 1,500 miles to reach our tables. This is completely unsustainable and depletes the nutritional value of produce. When fruits and vegetables grow, most of the vitamins and minerals are absorbed as they ripen on the mother plant. When we pick prematurely for shipping, their nutritional value is significantly compromised.

Local farmers are people we all should know. They are among the most important stewards of our precious home. Their critical role in our lives is always in the Symbiosis spotlight. They are the best purveyors of superior nourishment from The Earth. The family farmers have persevered, tending centuries of sacred knowledge and commitment, through unfathomable challenges. We must be proactive in support of small farmers as they push through these worst of times. We are buoyed by the renewed interest of the young people who seek to reclaim some control of our destiny. Here is an "other" teaching moment: we *are* the others. If we fail each other, we fail ourselves. Nowhere is that lesson clearer than our support of these special souls who provide the food for our children. Symbiosis sets the table so we can support our farmers and we can help them grow our next feast.

Symbiosis finds opportunities everywhere to thoroughly explore teaching moments as they evolve. It gives people direction to use what they've learned to help in their communities.

Symbiosis has animated features to entertain while they teach. We are all learning how "food production" is an absurdly complex issue. Our egos have led us to believe we are smarter than Nature. Now, we unnaturally inject genes and douse crops with toxic chemicals in a monoculture that destroys soil structure and water supplies. These issues including inhumane labor practices and the effects on our health, economy and food security are covered in the first Symbiosis animated short. Seth MacFarlane, Trey Parker and Matt Stone collaborate on this segment and they recruit more animators from their sphere and beyond. Many complex issues are brought home by their work. It is remarkable how serious points are accepted and actions are initiated when presented through skillful animation. Symbiosis has many viewer created animations that are available online and some are selected by consensus to be premiered on the series.

Food and gardens are a perfect segue to Symbiosis' first musical guest, Taja Sevelle who has been growing an organization since 2005 called Urban Farming. After she performs, she gives the audience a preview of next week's show when we will meet her in Detroit where she launched her first gardens.

On Symbiosis' visit to Detroit next week, we will pay homage to one of America's greatest scientists. Stan Ovshinsky was a self-educated genius inventor who held over 400 patents. The batteries and LCDs he invented are what make all our communication screens possible. He shared Charles Steinmetz's belief that science should always have a moral compass to serve humanity and share knowledge. Ovshinsky dedicated his life to make the world a better place and free us from fossil fuels with his inventions in solar and hydrogen technologies. Stan told us: "never stop going to your own school", he did just that until he left us in 2012 at 89 years young. We can only imagine how amazing things could be if we all shared knowledge freely. Something Maya Angelou always asked would apply: "what would this country be if we dared to be intelligent?" We will miss Stan but we will no doubt feel his spirit at his laboratory where his work lives on in other bright, spirited scientists. Detroit is now a hub of innovation rivaling California technology centers. We will bring a local eighth grade class to explore and see what remarkable things they are discovering. The kids will take away inspirations for science projects that Symbiosis will follow into the future.

We witness fields of food and solar panels planted where blocks of foreclosed houses recently stood rotting. We witness empowerment in bold beginnings of thriving, sustainable communities. Detroit is a place, despite its bad press, rising from the ashes. Symbiosis will have an eye on the flight of this rising Phoenix online and for future shows.

There are exceptional places where Symbiosis will revisit a few times. They are places abundant with teaching moments and play significant roles in the big picture. Such a place is an extraordinary fish farm, Veta La Palma in the south of Spain. There they practice extensive farming on a 27,000 acre farm with zero inputs and the fish feed themselves. It has naturally progressed to become sanctuary to thousands of birds. Some fly three hundred miles each day to dine at Veta La Palma. Their feast is an epiphany, these birds are a portent. What do we choose now for our children and for their tomorrows?

The farm acts as an enormous filter for the river that feeds it and is self-renewing. It produces tons of some of the cleanest fish on Earth and it is a profound fiscal triumph and a tribute to Nature herself.

The farm's truest success is derived from its strong bonds to nature. Miguel Medialdea, who enthusiastically takes us for a tour explains that he is not an expert on fish. He is an expert in relationships. He knows how all species must live together in harmony. This is a tremendous role model transferable to ALL kinds of farming and industries. Each visit to Veta La Palma fills us with inspirational ideas, there are several sites being restored like this today. Symbiosis will return, not just to raise consciousness about their existence, but to make them so familiar and popular that they become the attainable standard, bold operating systems that are inclusive of all life.

While in Spain, our Symbiosis Silence, Qi gong session and close of the show will explore incredible, stunning sections of the Camino de Santiago. There, we will experience the deep peace and beauty of this spiritual journey. This sacred walk is a wonderful collaboration of humans with Nature entwined by an intense, deep faith in both.

In an effort to walk the walk, from the first episode, Symbiosis will strive for carbon neutrality. Television has a ginormous footprint. While it will take some effort to be produced without any carbon footprint, we measured it from the start. We are transparent and vigilant about serious steps to minimize it. This is another effective method to raise the bar on our own Earth stewardship and has inspired others in the industry to follow our lead.

Symbiosis Web will be an all-encompassing resource center online. It provides specific information, contacts, addresses, maps, groups and businesses. Anyone can find out what they can do with career opportunities, organizations to join, actions they can take in their communities. Symbiosis Web will provide the connective threads to weave new fabrics for our Earth.

Symbiosis projects are accessed online for information, with connections to actual people and places. Many have webcams to monitor their progress. There are tutorials on most topics Symbiosis covers which are linked with interactive resources for in-depth open source learning.

Symbiosis is about rebuilding community connections. No one should be out there struggling unless it is by choice. When we share bonds with those around us, really know them, know their story, we, the individual will treat "Us" better and think before we act. That big question must always be asked: "How will what I do affect Us"? Perhaps revisit the Golden Rule?

It is Symbiosis' mission to change our course. In the same way that getting energy from caffeine and sugar will not sustain our bodies, we must stop ravaging The Earth for "cheap" fossil fuels. They are only cheap when we do not count the costs to every living thing (including ourselves) in our path to harvest and burn them. It does not include the daily struggle of people gasping for the fouled air. We especially ignore the powerless who are trapped in the shadows of refineries, hydro-fracturing and coal plants. They pay the supreme toll with their lives. Their children are born into an inescapable toxic stew in their air, water and soil. Those families, our fellow humans, will spend their lives battling asthma, cancers, neurological disorders and learning disabilities. Like the trees, sea and wildlife we ravage, these people are rendered invisible as we consume more. There is significant change emanating from Symbiosis, making a difference for generations lost to an apathetic system that had deemed their lives expendable. At Symbiosis we have adopted a crazy idea that life has infinitely more value than money. That one crazy vision, will save our home.

The sky has no limit. Teachers, historians, architects, engineers, inventors and others who are interested in rescuing our Earth from humans are embraced by Symbiosis. Some become household names whose words of wisdom are now embedded in our everyday conversations. Symbiosis is the opportune place for people of means, fame and power to join together. The stream of talent and resources will assemble in an unprecedented all-encompassing chance, to gain strength, and counter destructive forces who play dangerous roulette against Our Earth, against Our Only Home.

THIS teaching moment, will change perceptions and give clarity to our lives here on Earth. The outpouring is joyously contagious and humbling. Symbiosis wakes the unconscious and creates a better world with energy of all forms. Because everything is energy, people are energy, every being needs energy for life, continuing to destroy essential life to obtain energy, will mean the end of this life that we all share.

Symbiosis is the entity that changes the focus from what is wrong and what we can't do, to the grand things we do NOW and CAN do for our future. Most importantly, the good we can do for our children who deserve our best. Maybe, their future is more important than an SUV.

We could have saved The Earth

but we were too damned cheap

-Kurt Vonnegut

(If only we would prove Mr. Vonnegut wrong, this one time!)

EIGHTEEN

Origin of my quest and final thoughts:

Not everything that is faced, can be changed.

But nothing can be changed, until it is faced.

- James Baldwin

On the day after Thanksgiving 1960, on CBS, in prime time, America watched a one hour documentary by the incomparable Edward R. Murrow, titled "Harvest of Shame". It exposed harsh realities of how we treat the people who put the food on our tables. The force behind this unconscionable dehumanization were in these early seeds of corporate agriculture. America had been through this disgrace before, it was actual slavery. Only this time, the enslaved are legally free to leave, the chains are invisible but they are wrought from the rigid iron of social and economic reality. Fifty four years after we watched, our bellies stuffed with turkey, these workers remain shackled by an inhumane system. This system has been justified through time as necessary to feed a growing population, as the only way it can be done, though it lays waste to human beings and the Earth. As a further insult to humanity, global and local hunger issues are worse than ever, more a result of distribution and politics than production.

In 1972 (twelve years after Harvest of Shame), in eighth grade English, we studied "Of Mice and Men". My teacher (Mrs. Christoffel) expanded our lesson to cover Cesar Chavez and the struggles of present day Farm Workers. I was shocked at how little the world had changed since Steinbeck (in 1937) wrote of George and Lenny. Only the people I learned were toiling in America's fields were now whole families. It was incomprehensible and completely unacceptable to me that children, some younger than I was, were working in fields instead of being in school. The food on my table had passed through their little hands! As if on cue, a connection to the UFW appeared through a friend of a friend.

Besides sticking up for a couple of kids against some mean kids and a bit of tutoring, I had never stepped up to help. Now, with a clear opportunity and the motivation of egregious injustice, it was time for me to do something. I spent any free time in those formative years on picket lines and fund raising for La Causa. I have made a lifelong effort to leave anyplace I've been better than I found it, I consider it an obligation to life. I am indebted to my teacher for her inspiration to take this path so rich in life lessons and witness the strength and courage of so many.

Now, over forty years after I stepped up, conditions for those who harvest food in America and we now import from other nations, are still unconscionable and getting worse. Why do we continue to devalue these people who provide our food? Where we produce this food, we destroy topsoil, pollute and deplete the water. In California, we turned desert into farmland by stealing water from the mountains and the people south of the state. Recently, depletion of the Sierra Nevada snowpack is putting an end to that theft with a prolonged drought. In other places, like Peru and Indonesia, our insatiable demands create new deserts and destroy rain forests, decimating other species. We ignore Nature at our own peril.

We the people, who know the truth, are on the rise. We are stoked in optimism every day by extraordinary individuals, joyfully and tirelessly united. We are seedlings, reaching toward the light, growing together to become a resilient stand of oaks. We, can create the difference. We, shout from the mountains, from precarious perches and rooftops. We, need a clear voice and image, and WE, all the people, must be seen on a world stage.

Symbiosis is that stage, that clear voice, clear with knowledge and ideas to embrace our children. All ideas, new and old, are welcomed with one crucial requisite: *every action must support all life*. We must refocus on Nature as our new sage, an omniscient mentor to guide our path. The Earth is not our adversary, she provides all that is necessary, all the comforts of home.

If one is lucky,

one can pass that comfort on to a child,

in peace and with love.

Many thanks to you for reading.

NINETEEN

There Will Come Soft Rains

There will come soft rains and the smell of the ground,

And the swallows circling with their shimmering sound;

And frogs in the pool singing at night,

And wild plum trees in tremulous white;

Robins will wear their feathery fire,

Whistling their whims on a low fence wire;

And not one will know of the war, not one

Will care at last when it is done.

No none would mind, neither bird nor tree,

If mankind perished utterly:

And Spring herself when she woke at dawn

Would scarcely know that we were gone.

-Sara Teasdale

(from Ray Bradbury's story of same title)

The moment
we cease to hold each other,
the sea engulfs us
and the light goes out.

- James Baldwin

TWENTY

Got a mind full of questions
and a teacher in my soul,
and so it goes…
-Eddie Vedder

By third grade, I knew in my heart, I was an artist. To this day it is deep in my core. By eighth grade the activist in me was ignited and obligated me to always work toward fairness. I became a worker bee, climbed a corporate ladder and by chance, found myself an entrepreneur. I grew up on a suburban island in New York and was soon lured to the urban island of Manhattan. Island life, even on those islands "developed" by man, presents many unique perspectives. On islands, Nature's laws are in complete focus, respect is quickly taught and human control is exposed as a profound illusion.

On my journey, I've met hard challenges that make me grateful for the luck of second chances offered by the universe. These days, I spend much of my time on a rural island in the Canadian Maritimes. There, I advocate and pioneer the building of beautiful homes with non-toxic, carbon capturing cellulosic masonry. Their walls naturally minimize the need for a heating or cooling energy source. With a carbon footprint of zero, they will easily exist in harmony with The Earth for hundreds of years. These structures, after providing warm shelter are easily returned as nourishment for the soil. If we all learn to consider The Earth in what we leave behind, our descendants will be able to choose to live in these homes, upcycle them or return them to nourish The Earth. That's a choice I would love to leave for those I leave behind, as my best contribution toward their future.

Fail to teach, we lose our virtue. Fail to share, we lose our humanity.
Learning is essential, your ideas and thoughts are appreciated:

Valerie Bline
446 Willis Ave Suite 190
Williston Park, NY
11596

valeriebbline@gmail.com

www.ingramcontent.com/pod-product-compliance
Lightning Source LLC
Chambersburg PA
CBHW070948210326
41520CB00021B/7107